Quick and Easy Stirling Engine

By Jim R. Larsen

Published by: Jim R. Larsen, P.O. Box 813, Olympia, WA 98507

ISBN-13: 978-1466277779

ISBN-10: 1466277777

Revised 2/1/2012 – Minor formatting changes and corrections.

About the Author

Jim R. Larsen is a big fan of the Stirling Cycle Engine, and the author of other popular Stirling engine titles, including "Three LTD Stirling Engines You Can Build Without a Machine Shop", and "Eleven Stirling Engine Projects You Can Build". He has been designing and building model Stirling Engines for many years. Jim is a teacher, a trainer, an author, and an artist. He strives to design and make models that are unique, fascinating, and that you can build yourself.

Contacting Jim R. Larsen

YouTube
Jim periodically posts videos of Stirling engines on YouTube. His YouTube username is 16Strings. The web address for Jim's YouTube channel is **http://www.youtube.com/16strings**.

Web Page
Visit **http://Stirlingbuilder.com**.

Email
You can send email to Jim at **Jim@Stirlingbuilder.com**.

Jim's Blog
Visit **http://woodenmusic.blogspot.com/**

Work Safe! The publisher takes no responsibility for the use of any of the materials or methods described in this book, or for the products thereof. Always use power tools and hand tools in a safe manner.

Table of Contents

Introduction

The Quick and Easy Stirling Engine can be assembled from soda cans with simple hand tools in as little as three hours. This engine was designed by Jim Larsen for use in educational settings with students young and old.

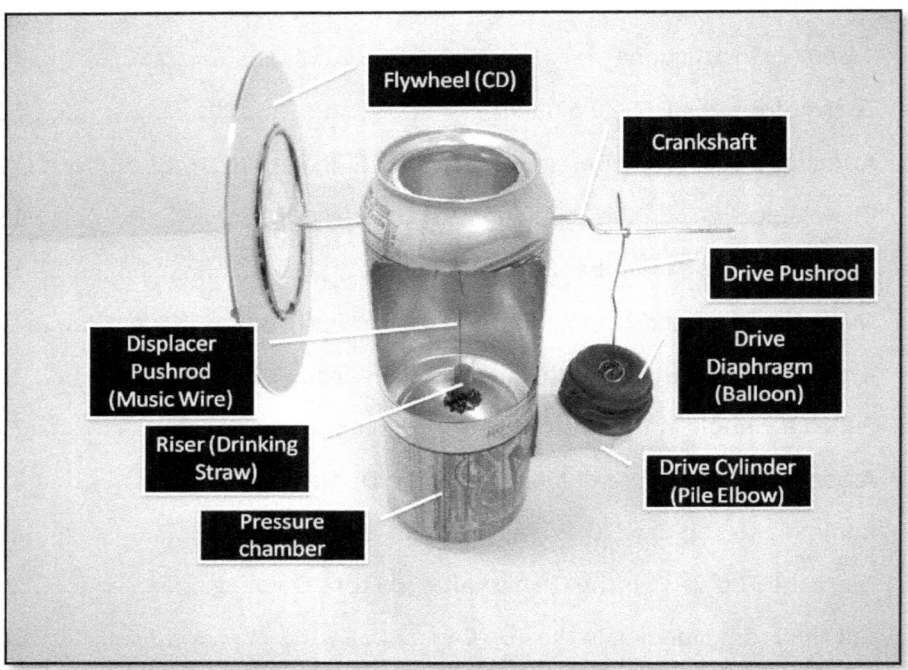

Figure 1 - Quick and Easy Stirling Engine

Tools and Materials Needed

- 2 (or more) empty 12 ounce aluminum soda cans
- 1 unopened 12 ounce aluminum soda can
- 1 medium size helium quality latex balloon
- 3/4" (19 mm) x 1/2" (13 mm) PVC pipe elbow
- 1 piece of 0.015" (0.38 mm) music wire (or guitar string) at least 10" (254 mm) long
- 16" (406 mm) copper wire, 12 or 14 gauge
- J-B Kwik® fast setting epoxy
- Scissors

- Pliers with wire cutters
- Paper hole punch
- Heat gun
- Gloves (for heat protection)
- Permanent marker
- Sand paper
- Small pin or needle
- Floral arrangement foam (one 4" x 9" "brick" makes 8 engines)
- Large diameter drinking straw
- Paper and pencil
- 1 CD
- Ruler
- Small coin (US dime or equivalent)
- Tape
- Super Glue®
- 19 gauge steel wire (or craft wire)

If you are working as part of a group, many of the supplies can be shared. Things like music wire, drinking straws, and floral foam provide enough material to be used in several motors.

About the Materials
All the **soda cans** used should be of the same size. The brand name does not matter, just as long as the cans are the same diameter. If you are using European soda cans, they will work in the same manner described here.

The helium quality **balloons** can be found in most any toy department.

The **PVC pipe elbow** is available in the plumbing department at most any hardware store. It does not matter if the inside of the fitting is smooth or threaded. If there is raised lettering or mold marks on the outside of the fitting they may need to be removed with sand paper to prevent the balloon diaphragm from leaking.

0.015" **music wire** can be purchased at model shops (where radio controlled models are sold) and some hardware stores for about 20 cents for a 36" piece. Smooth (non-wound) guitar strings are about the same size and will also work.

Copper wire was chosen for the crankshaft because it is easy to bend, yet stiff enough to hold its shape under these light loads. It can be purchased in the electrical department at hardware stores. It is also possible to use coat hanger wire for crankshaft construction. It is harder to bend, but holds its shape better due to increased stiffness.

J-B Kwik® epoxy was chosen for this project for two important reasons. It cures in a matter of minutes, and it is able to withstand temperatures up to 300° F. If you substitute a different adhesive, check the temperature rating to make sure it can tolerate at least 300° F.

Small **scissors** with a sharp point work very well for cutting aluminum cans. Medium sized scissors work well for the final straight cuts.

The **heat gun** is perhaps the least common of the tools in this list. They are not expensive. If this engine is being built as a project in a class, the entire class should be able to share the use of a single heat gun. The same is true for the **gloves**.

Floral arrangement foam is available in craft stores or online under a variety of product names. It is usually green, and often sold in blocks that measure about 3" (76 mm) x 3" (76 mm) x 8" (203 mm). One block will make displacers for 7 or 8 engines. The foam will specify if it is for dry flowers or for live (wet) flowers. Either product will work. This product was chosen because it does not melt when it gets hot, but it will burn, so be careful around open flames. It is very easy to shape and form, and very light weight.

The **drinking straw** is the type commonly used when serving sodas at the local hamburger establishment. Larger diameter drinking straws are preferred for this application because it makes them easier to install. There is sometimes a need for a small diameter coffee straw, but only for some troubleshooting applications.

The **CD** can be new or used, blank or with media. It can have a label or not. Just don't use anything that you ever want to use as a CD again in the future. Once it becomes a flywheel, it is rendered useless from its previous life.

The **19 gauge steel wire** (or craft wire) is used to make the pushrod for the drive diaphragm. It is light gauge steel wire, similar to what is used in floral arrangements. The music wire or copper wire will work as a substitute.

How Hot Air Engines Work

The pressure chamber contains a small amount of air that is held captive inside the engine. One end of the engine is warm, and the other end is cool. The displacer moves the air inside the engine back and forth repeatedly, from warm, to cold, and to warm again.

The air inside the engine expands when it gets warm, and pushes outward on the drive mechanism. When this same air is moved to the cool side of the engine, it contracts. This pulls in on the drive mechanism.

The drive mechanism pushes and pulls on the crankshaft. This causes the crankshaft and flywheel to rotate. The rotation of the crankshaft causes the displacer to rise and fall inside the pressure chamber.

The rotating crankshaft causes the cycle to repeat. The air heats, expands, and pushes the crankshaft through the expansion phase. This moves the displacer and causes the air to enter the cool side of the engine. The air then cools, contracts, and pulls the crankshaft through the contraction phase. This starts the next expansion phase and the pattern continues to repeat itself.

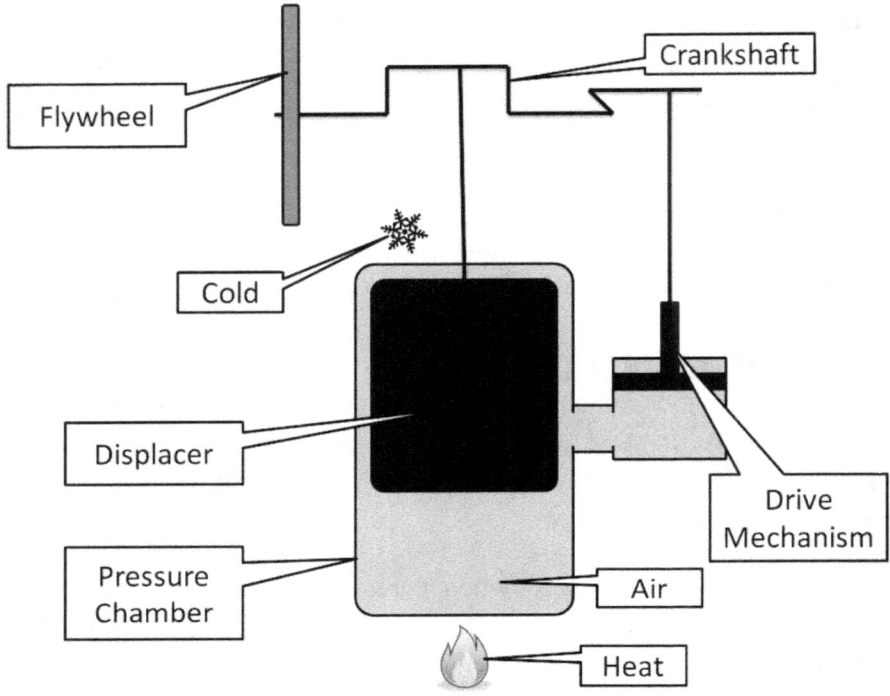

Figure 2 - Stirling Engine Parts

This drawing illustrates the basic parts of a Stirling engine. The pressure chamber contains air that is held captive in a closed system. The air is heated when it is on the bottom of the pressure chamber, near the flame. The air is cooled when it is on the top of the pressure chamber. The air is moved about inside the pressure chamber by a loose fitting piston called a displacer.

The drive mechanism pictured here is a piston. The drive mechanism you will be building uses a rubber diaphragm. They both accomplish the same thing.

The crankshaft and the flywheel keep everything moving in sync.

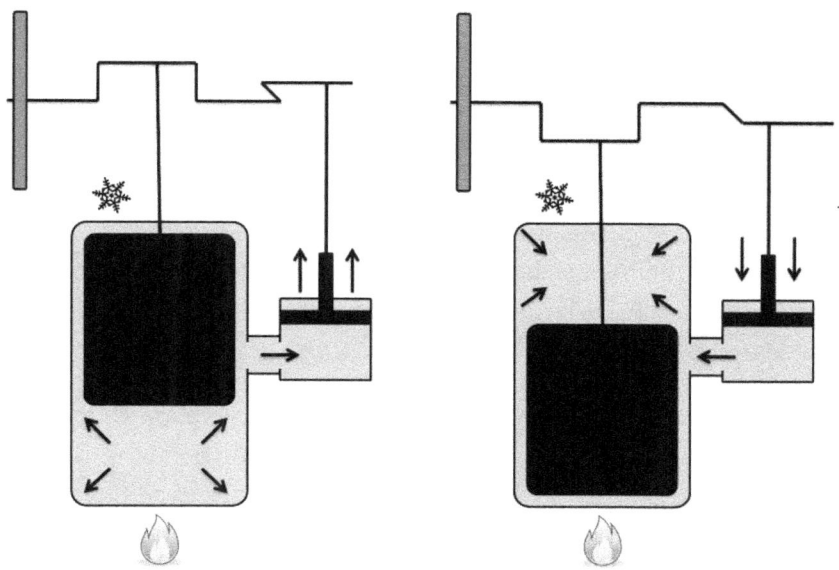

Figure 3 - Warming Phase   **Figure 4 - Cooling Phase**

When the engine is in the warming phase, the air is in the warm side of the engine. This causes the air to expand, which pushes the drive mechanism upward.

When the engine is in the cooling phase, the air is on the cool side of the pressure chamber. The air contracts as it cools. This pulls down on the drive mechanism. The repeated pushing and pulling of the drive mechanism causes the crankshaft and flywheel to rotate.

Assembly Instructions

Thermoform the Drive Cylinder

The drive cylinder is fashioned from a 3/4" (19 mm) x 1/2" (13 mm) 90° PVC pipe elbow. The wide end of the pipe elbow will hold a balloon diaphragm. The narrow end of the pipe elbow will be attached to the side of the engine. The attachment to the round side of the can is made easier by forming the pipe elbow to match the curve of the can. This makes it easy to attach to the can with epoxy and maintain a strong joint with a good seal.

There are several ways to shape the end of the pipe elbow to match the curve of the can. If you do not have the ability to heat and bend the part as illustrated here, you can also shape it with sandpaper or a curved file. Sanding and filing takes a lot longer and makes a dusty mess, so the preferred method is thermoforming.

A heat gun set to "low" seems to work very well as a heat source for thermoforming the PVC pipe fitting. (Testing with a propane torch showed that the propane flame is too hot for this process.)

Tools needed for this step:

- Heat gun (or other heat source)
- 3/4" (19 mm) x 1/2" (13 mm) PVC pipe elbow
- Heavy glove (to protect your hand from heat)
- 1 unopened can of soda

Wear a heavy glove to protect your hand from the heat. Warm the small end of the PVC pipe elbow with the heat gun (or similar heat source) until the plastic is easy to bend with slight pressure. Press the warm soft plastic firmly against the unopened soda can so that the end of the pipe elbow bends to take on the curve of the can. Hold the pipe elbow against the soda can until the plastic cools and holds its new shape.

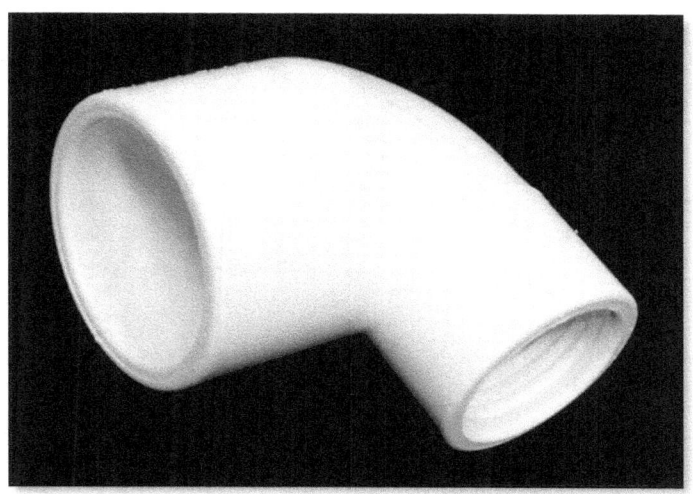

Figure 5 - The drive cylinder is made from a PVC pipe elbow.

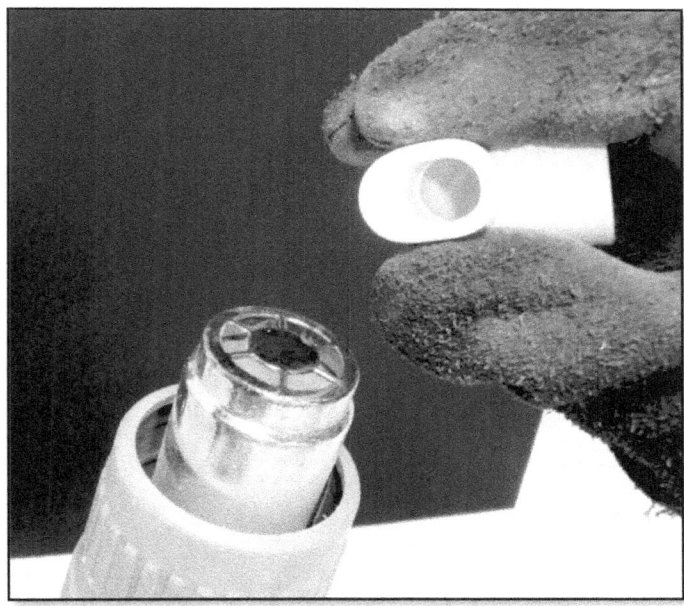

Figure 6 - Heat the small end until it becomes soft. You can tell when it is getting soft as it will deform easily when pressure is applied, as you can see here.

Figure 7 - Press the heated end against an unopened soda can and hold it until the fitting cools. The PVC fitting will take on the contour of the outside of the can.

Attach the Balloon Diaphragm

Tools needed for this step:

- 1 medium sized helium quality balloon
- Scissors

Cut the neck from the balloon as illustrated in the pictures. The middle section of the balloon can be discarded. It is not needed for this project.

Stretch the top of the balloon over the large end of the pipe elbow.

Roll the neck of the balloon into a donut shaped rubber band and stretch it over the outside of the balloon to hold the balloon in place on the pipe elbow.

Carefully remove any wrinkles in the balloon fabric at the point where the rubber band is holding it against the pipe elbow. Wrinkles may cause leaks later, and leaks are not good.

Adjust the diaphragm so that it has enough slack to pop in or out of the end of the pipe elbow. There should be enough slack in the balloon to allow the center of the balloon to move in or out about 1/4" (6 mm) without needing to stretch the balloon.

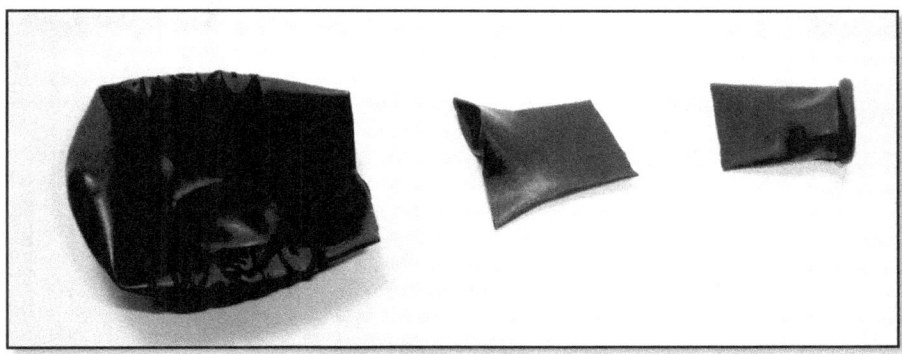

Figure 8 - The drive diaphragm is made from a helium balloon. The balloon top (left) will become the drive diaphragm. The balloon neck (right) will be rolled into a band and used to secure the drive diaphragm to the pipe fitting. The center section of the balloon may be discarded.

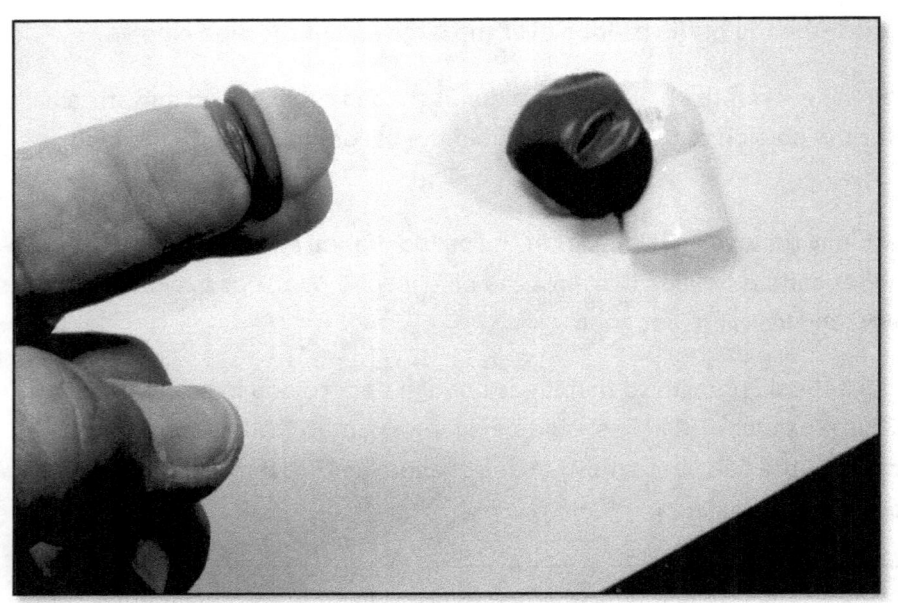

Figure 9 - Roll the balloon neck to form a thick band of rubber.

Figure 10 - Use the rubber band made from the balloon neck to hold the diaphragm in place.

Draw Guidelines on the Soda Cans

Tools needed for this step:

- 2 empty soda cans
- 1 fine tipped permanent marker

Pick two empty soda cans and designate one as the engine bottom and one as the engine top.

Bottom Can:

Draw a line on the side of the *bottom can* that is 2 1/2" (64 mm) up from the bottom edge of the can. Ignore the curvy bottom section of the can. Measure only the flat sidewall section of the can. You need 2 1/2" (64 mm) of straight sidewall.

The easy way to mark the can is to hold the marking pen steady at the proper height above a flat surface. Use a stack of books or magazines to support the pen at the correct height. Hold the can against the pen and rotate the can to draw the line around the can.

Top Can:

Measure and mark a line around the *top can* that is 1/8" (3 mm) above the bottom edge of the sidewall. Just like before, ignore the curvy bottom part of the can and mark the line 1/8" (3 mm) up the sidewall.

Figure 11 - The bottom can (left) is marked at 2 1/2" (64 mm) from the bottom. The top can (right) is marked 1/8" (3 mm) from the bottom edge. The brand of beverage is not important, but both cans must be the same size.

Cut the Pressure Chamber Bottom Can

You will want a clean straight cut along the line drawn on the bottom can. The best way to get a smooth clean cut is to make the cut in a two step process. Make an initial cut about 1/4" (6 mm) above the line and discard the top of the can. This cut is likely to be a bit rough because the aluminum tends to wrinkle as you cut it. Make the second cut carefully right on the line. This second cut is much easier to make straight and even because the top of the can is now out of the way.

Figure 12 - Make the initial rough cut about 1/4" (6 mm) above the line.

Figure 13 - Make the final smooth cut right on the line.

17

Make the Displacer Piston and the Displacer Pushrod

Tools needed for this step:

- Floral foam (at least 3" (76 mm) x 3" (76 mm) x 1" (25 mm))
- Ruler
- Marker
- Music wire (or a guitar string)
- Wire cutters
- J-B Kwik® Epoxy
- Bottom can

Floral foam is soft foam used for holding flower arrangements. Some varieties are made for holding wet flowers, others are specifically for dry flowers. Either product will work. It is the preferred material because it is easy to cut and form, and it tolerates the heat well.

Mark the end of a floral foam brick and slice off a section exactly 1" (25 mm) thick.

Use the bottom can like a cookie cutter and punch a disk out of the middle of the 1" (25 mm) foam slice.

If the foam is stuck inside the can, don't worry. The foam disk needs to be slightly smaller than the inside diameter of the can. You can begin to make it smaller now by rolling the can in your hands and applying pressure to the outside of the can. The foam is easy to crush. Be careful so that your crushing is controlled and has the desired effect.

Remove the foam from the end of the can and continue to roll it on a firm surface until it is small enough to drop inside the can and fall under its own weight. You can also rub the foam with your fingers to remove excess loose material. When the disk will drop inside the can under its own weight, stop making it smaller. It is now the correct diameter.

If this process messed up the bottom can, make another one. It only takes a couple of minutes and you don't want to work with a messed up or bent can.

Cut a piece of .015" (0.38 mm) music wire to a length of 10" (254 mm). Music wire can be purchased at model shops and some hardware stores. If you can't find any, use a guitar string. The B string or high E string is about the right size.

Make a 90° bend about 1" (25 mm) from one end of the music wire.

Insert the long end of the music wire through the center of the foam disk. Look carefully to make sure it is as close to the center as you think you can get it, and as close to "straight through" the disk as possible. If it looks crooked or off center, pull it out and try again. The floral foam will tolerate several attempts at inserting the wire.

Use J-B Kwik® epoxy to attach the foam disk to the 1" (25 mm) bent portion of the music wire.

Figure 14 – Cut a 1" (25 mm) slab from the end of a block of floral foam.

19

Figure 15 - Use the pressure chamber bottom like a cookie cutter to create the displacer disk.

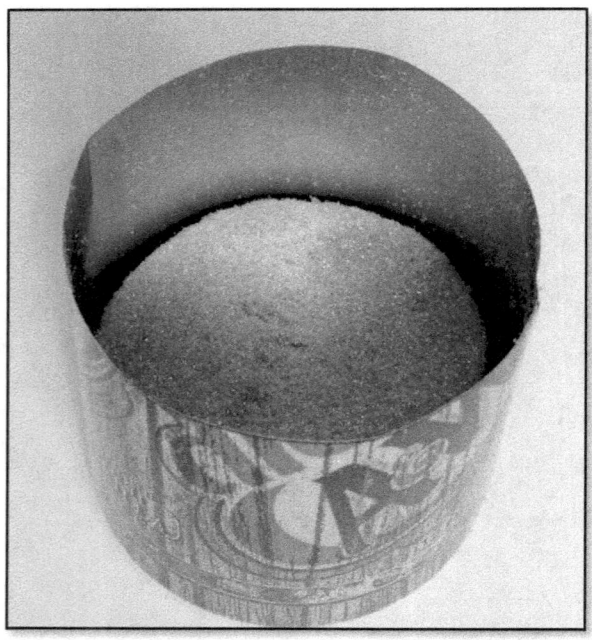

Figure 16 - The displacer disk is reduced in size until it will fall into the soda can under its own weight.

Figure 17 - The music wire is passed through the foam disk and glued with J-B Kwik® epoxy.

Make a Hole for the Drive Cylinder Attachment

Use a paper punch to make a hole in the side of the bottom can. This is to allow air to flow in and out of the drive cylinder (pipe elbow). The drive cylinder will be attached over this hole. Make the hole about 1" (25 mm) from the edge of the can (which is about as far as you can reach with most paper punches).

If you don't have a paper punch, then use a sharp object or some small pointy scissors to cut the hole. Flatten any rough edges around the hole so that they don't get in the way of the displacer's motion inside the can.

Figure 18 - Use a paper punch or a sharp cutting tool to make a hole in the side of the pressure chamber. The drive cylinder will attach here.

Punch a Pinhole in the Bottom of the Top Can

Tools needed for this step:

- Small pin or sewing needle
- Pliers
- Top Can

A pinhole must be made in the exact center of the bottom of the can that is designated as the "top can" of the engine. This pinhole must be large enough to let the music wire pass through, but not much bigger. **If the hole is too big, pressure will leak and the engine will not run.**

Grasp a small pin or sewing needle with pliers and carefully poke a hole in the bottom center of the top can. Test the fit with the music wire. If the hole seems to be larger than the wire, pick another can and try again. This step is important!

Assemble the Pressure Chamber with the Displacer

Thread the music wire connecting rod up through the bottom of the top can, and then place that assembly over the bottom can, with the displacer now inside the bottom can.

Carefully press the top can into the bottom can. The bottom can will stretch and expand to allow the top can to be pressed in. This is what the 1/8" (3 mm) mark is for on the bottom of the top can. Press the can in until the line is even with the top edge of the bottom can.

Figure 19 - Pass the music wire through the hole in the bottom of the can, then press this can into the pressure chamber bottom, up to the line.

Figure 20 - The top can is pressed down into the pressure chamber bottom until the line on the top can is lined up with the seam.

25

Remove the Top Can Lid

Use a pair of scissors to cut away the inside of the top can lid. Leave the rim around the top of the can in place. This adds strength to the can. Some people prefer to remove the can lid before assembling the parts together, and that is fine too. The lid adds some structural stability to the can and makes it easier to press it into the lower can without bending or distorting the sides of the upper can.

Figure 21 - Cut away the inside of the can lid, keeping the rim for strength.

Mark the Top Can for the Crankshaft and Front Opening

The crankshaft will be mounted directly above the hole in the side of the bottom can. Use a straight edge and a marking pen to draw a line from the hole to the top of the engine. Draw another vertical line on the opposite side of the engine. These will act as the centerline of the motor, and will be used to position the crankshaft holes.

Draw a line around the circumference of the top can that is 3/4" (19 mm) above the seam (where the two cans join).

Draw a line around the circumference of the top can that is about 1/8" (3 mm) below the top edge of the sidewall of the top can.

Draw vertical lines parallel to the centerlines, about 1/2" (13 mm) to each side of the centerline.

Need Help?
Visit http://StirlingBuilder.com to view video tutorials showing how to construct the Quick and Easy Stirling Engine.

Figure 22 - Draw a centerline up from the hole. Draw parallel lines 1/2" (13 mm) from the center line. Draw a line 3/4" (19 mm) above the seam and 1/8" (3 mm) below the top edge. Repeat the pattern on the opposite side of the can.

Cut the Front Opening in the Top Can

The marks made in the previous step are the guidelines for cutting the opening in the front of the upper can. The top line around the can is the top of the hole, and the bottom of the hole is 3/4" (19 mm) above the seam. The sides of the hole are on the lines drawn 1/2" (13 mm) from the engine centerline. The corners of the opening are rounded to about the same radius as a US quarter. The rounded corners will make the hole easier to cut and help maintain rigidity in the sidewall of the can.

Carefully cut the access opening in the front of the top can.

Carefully punch a hole on each side of the engine for the crankshaft. The holes will be punched at the point where the centerlines cross the top mark, 1/8" (3 mm) below the top edge of the can.

Figure 23 - Carefully cut the opening in the side of the top can.

Figure 24 - Punch a hole for the axle (crankshaft) on both sides of the engine at the point where the centerline crosses the top line. The hole should be the same size as the crankshaft material.

Install a Drinking Straw Riser Around the Pinhole and Pushrod

Cool water or ice will be added to the top of the engine to help it run. Installing a small riser around the center hole will help prevent water from getting inside the motor and will let you add more ice or water than you would if you don't add this part. It is a bit of a challenge to glue this in place without getting any glue on the music wire or the pinhole. This step is optional, but is highly recommended.

Tools needed for this step:

- Large diameter drinking straw (Starbucks® cold drink straws work well)
- J-B Kwik® epoxy
- Scissors

Cut a 3/4" (19 mm) length of large diameter drinking straw. Place it around the music wire and let it rest against the top of the can. Very carefully glue the straw in place with J-B Kwik® epoxy around the outside of the straw. Do not let any glue get in the pinhole or on the music wire. Spread the glue all around the outside of the straw so that it forms a water tight seal.

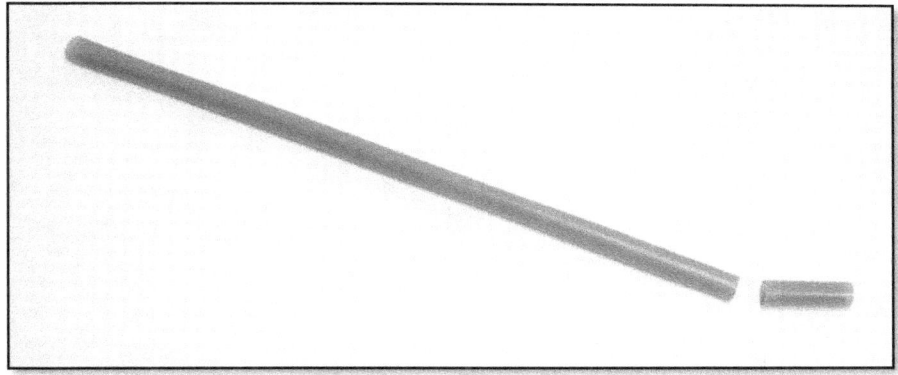

Figure 25 - Cut a 3/4" piece from the end of a large diameter drinking straw.

Figure 26 - Glue the drinking straw in place around the pinhole and music wire. Be very careful that you do not get any glue on the music wire or the pinhole.

Attach the Drive Cylinder/Diaphragm Assembly

Tools needed for this step:

- Sandpaper
- J-B Kwik® epoxy

Locate the pipe elbow and balloon diaphragm that was assembled previously.

Use sandpaper to roughen the surface of the can around the hole. This will help the epoxy stick to the can.

Glue the pipe elbow onto the can with J-B Kwik® epoxy. The balloon diaphragm must be facing up.

Figure 27 - Apply J-B Kwik® epoxy to the small shaped end of the PVC pipe elbow and glue it over the hole in the side of the pressure chamber.

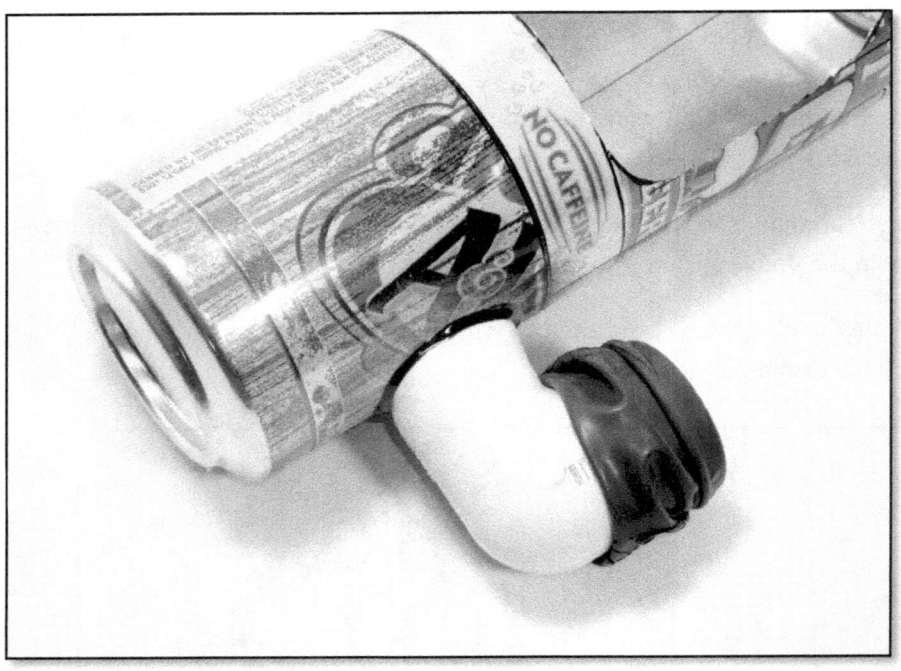

Figure 28 - The pipe elbow has been glued in place over the hole in the pressure chamber. Use sandpaper to roughen the side of the can prior to gluing.

Bend the Crankshaft and Attach the CD Flywheel

The crankshaft is one of the more difficult pieces to make for this engine. The process is made easier by drawing a template to guide you in bending the crankshaft. It is also helpful to remember to always keep the main parts of the crankshaft parallel to the centerline of the crankshaft.

Cut a piece of wire

The crankshaft for this project is made from 14 gauge copper electrical wire that has had the insulation stripped off. 12 gauge copper wire also works well. It can be purchased in the hardware store, with or without the insulating vinyl covering. It is also possible to use other stiff wire, such as the wire from a metal coat hanger. Start with a piece of wire that is 16" (406 mm) in length.

Draw the Crankshaft Template

1. Draw a straight line lengthwise down the center of a piece of plain white paper.
2. Measure and draw a parallel line exactly 3/8" (10 mm) from the centerline.
3. Measure and draw a second parallel line exactly 3/16" (5 mm) away from the other side of the center line.

The two lines that are 3/16" (5 mm) apart are the guide for bending the "drive section" of the crankshaft. This is the part that is over the drive diaphragm.

The two lines that are 3/8" (10 mm) apart are the guide for bending the "displacer section" of the crankshaft. This is the part in the center of the engine that lifts the displacer.

VERY IMPORTANT: The "displacer section" and the "drive section" of the crankshaft are offset by 90° of rotation. As the crankshaft rotates, the displacer section is 90° ahead of the drive section. When the displacer section of the crankshaft is at the top of its rotation, the drive section will be halfway up.

Bend the Drive Section

Begin bending from the "drive section" end of the crankshaft. Make the first two bends so that there is a 3/16" (5 mm) offset between the two straight sections. Use the template to make sure the long straight sections are parallel. The "drive section" should be about 2" (51 mm) long at this point. You may trim off any excess length later.

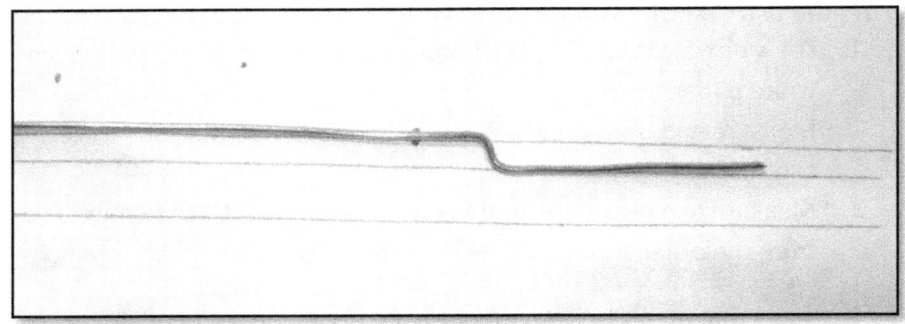

Figure 29 - The first leg of the crankshaft is for the drive section. It has an offset of 3/16" (5 mm).

Bend the Displacer Section

Hold the crankshaft in place on the motor and determine the best location for the bends of the "displacer section".

The section of the crankshaft that lifts the displacer will have a 3/8" (10 mm) offset from center.

Use the lines on the guide to measure the distance and to help you keep the main sections of the crankshaft parallel as you make it. Rotate the crankshaft to make sure that all the straight sections remain parallel to the axis of the flywheel.

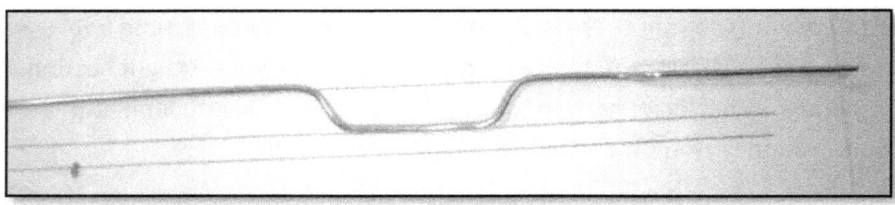

Figure 30 - The middle section of the crankshaft is for the displacer section. It has an offset of 3/8" (10 mm).

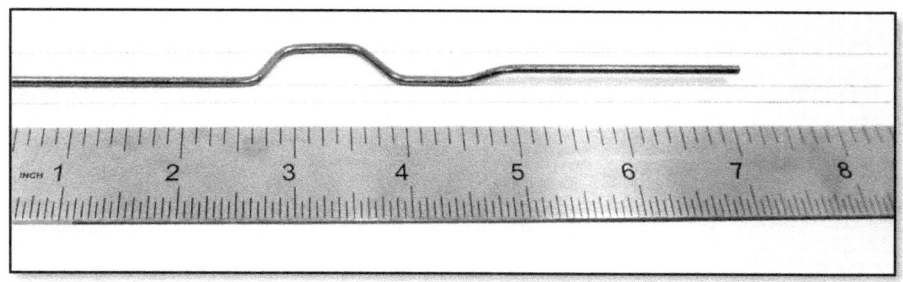

Figure 31 - This is a similar crankshaft. Note how the two sections are offset by 90°.

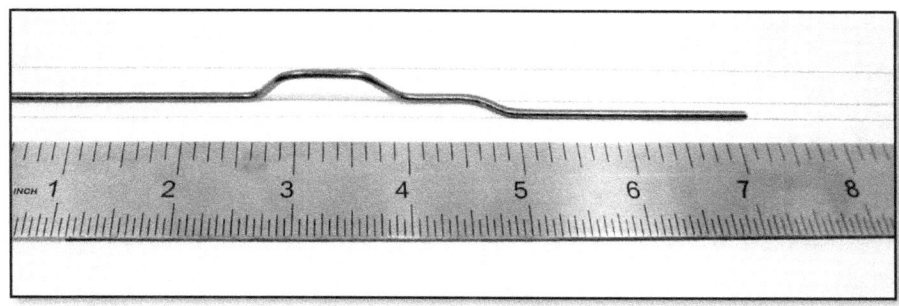

Figure 32 – Here is another view of the same crankshaft, showing how the two sections are offset by 90°.

You will notice that each bend in the crankshaft in these illustrations is about 45°. This makes it possible to adjust the offset distance of each section of the crankshaft by changing the angle of this bend. This allows you to fine-tune the crankshaft as you make it. Note that the crankshaft lines up with the centerline on both sides of the displacer section. The 45° bends simplify the installation of the crankshaft in the soda can.

Finally, make a loop for attaching the flywheel. The loop can be made by wrapping the wire around a soda can, and then adding a couple of bends so that the loop is centered around the center point of the crankshaft. Attach a CD to the loop with J-B Kwik® epoxy.

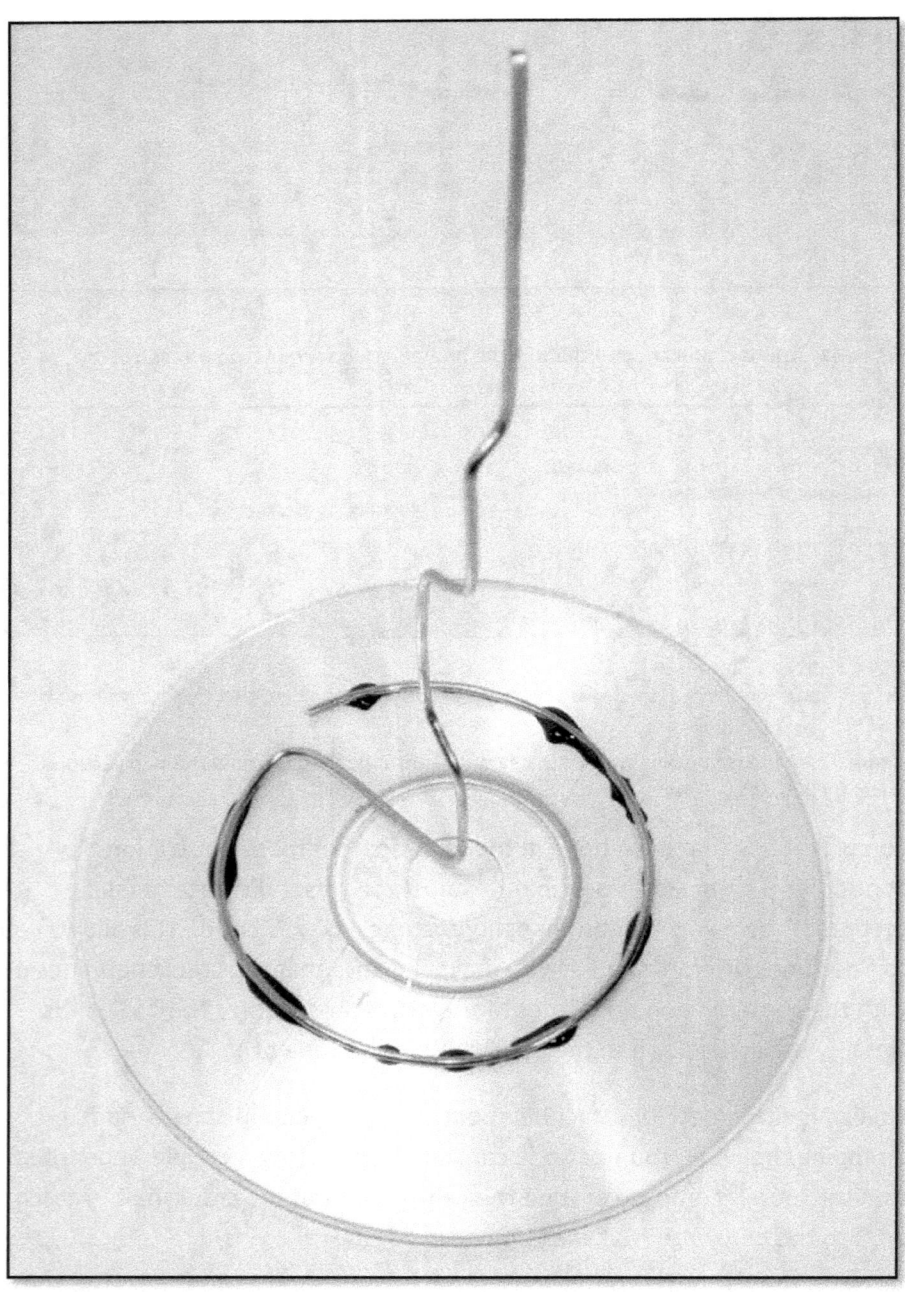

Figure 33 - This is the crankshaft with the flywheel attached. Note how the loop is formed in the end of the wire to attach the flywheel.

38

Install the Crankshaft and Attach the Pushrod

Carefully install the crankshaft and test the rotation. Check the "phase angle" to make sure that the drive section and the displacer section are 90° apart in rotation.

The music wire may now be attached to the crankshaft. Make a Z-shaped bend about 1" (25 mm) above the straw riser. This bend will allow you to adjust the length of the pushrod later, as you fine tune your engine.

The displacer must be attached to the crankshaft so that it will not be obstructed as it moves up and down inside the pressure chamber. It must not hit the top or the bottom of the pressure chamber as the crankshaft rotates.

Place the displacer section of the crankshaft in the "down" position. Place the displacer in the "down" position. Bend the music wire around the displacer section of the crankshaft and rotate the crankshaft to test the fit. Minor changes in length can be made by changing the angle of the z-shaped bend in the music wire.

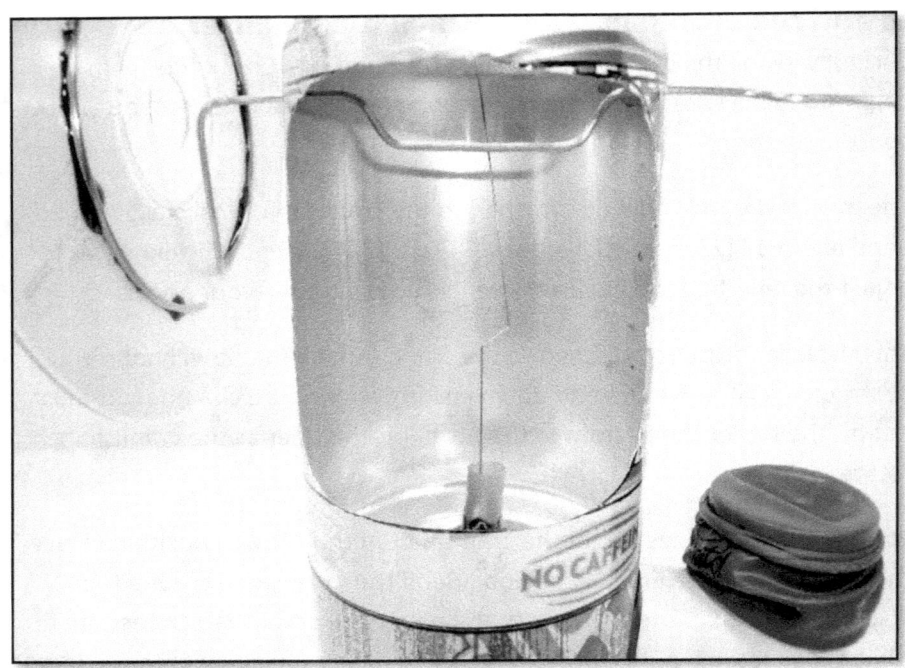

Figure 34 - There is a Z-shaped bend in the displacer pushrod. This makes it possible to adjust the length of the pushrod after it is attached to the crankshaft. Attach it to the crankshaft with at least one half turn around the shaft.

Balance the Flywheel

The flywheel needs to be tested for balance. This is easiest to do before the drive diaphragm is attached. Rotate the flywheel and let it coast to a stop several times. It is out of balance if it always stops in the same position, especially if it swings back and forth as it settles into that position. If it does not appear to be balanced, tape a small object, such as a dime or a small coin, to the CD. Manipulate the position of the counterweight until the flywheel appears to be balanced. A small coin, such as a US dime, is usually all that is required. Moving the weight towards the outside of the flywheel increases the effect of the weight. Moving the weight towards the center of the flywheel decreases the effect of the weight.

Figure 35 - A small coin is taped to the flywheel to provide balance.

Create and Attach the Drive Pushrod

A small wire pushrod will be made to connect the balloon diaphragm to the drive section of the crankshaft. Use 19 gauge steel wire or a similar material. If you don't have access to 19 gauge steel wire, you can use a piece of music wire, a piece of copper wire, or craft wire.

Begin with a piece of wire about 6" long. Make a loop in one end about 1/2" (13 mm) in diameter. Bend the wire so that this loop rests flush against the surface of the balloon diaphragm.

Make a Z-shaped bend about 1" (25 mm) above the loop. This bend can be adjusted later to change the length of the pushrod.

Hold the pushrod in position on the motor to determine the position for the attachment point to the crankshaft. The end of the pushrod should be situated so that the balloon diaphragm travels an equal distance up out of the pipe elbow and down into the pipe elbow. There should be just enough slack in the balloon fabric to allow this motion to occur without having to stretch the balloon.

Once the ideal position is determined, attach the pushrod to the crankshaft by wrapping it once around the drive section of the crankshaft.

Attach the loop end of the pushrod to the balloon diaphragm with Super Glue®.

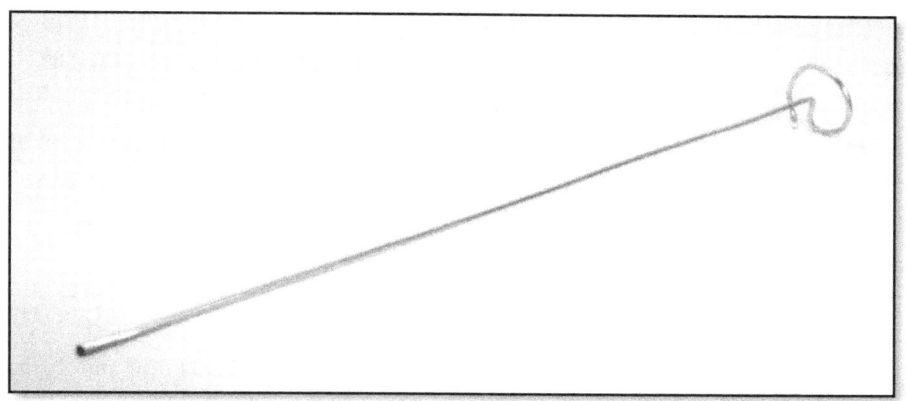

Figure 36 - This is the drive pushrod. The loop will be attached to the balloon with Super Glue®. A "Z" shaped bend will be added about 1" (25 mm) above the loop.

![Figure 37 image]

Figure 37 - The pushrod is attached to the balloon with Super Glue. There is a Z-shaped bend in the middle of the rod for adjusting the length (It is difficult to see at this angle). It is attached to the crankshaft with a single wrap around the shaft.

Preflight Checklist

You should now have a fully assembled Stirling engine that is ready for its first running. These things almost always need some tweaking and adjustment before they will run continuously, so don't be alarmed if it takes a little effort to get it working the first time. After you get it adjusted and "tuned" it will be much easier to start and run.

The pressure chamber must have no leaks! Every seam in the pressure chamber is a possible leak. The joint where the two cans are pressed together must be air tight. The pinhole must be a close fit around the displacer pushrod. The pipe elbow must be completely sealed where it joins the engine body. The diaphragm must seal tightly around the top of the pipe elbow.

Any friction that can be eliminated must be eliminated! Even things that look like only a tiny bit of friction must be dealt with if at all possible. These engines have just enough power to overcome their own friction and make themselves go. It takes very little to slow them down or make them stop. Friction happens everywhere that two moving parts touch each other, and every place where a moving part encounters the atmosphere. The crankshaft must turn freely without binding. The wires connected to the crankshaft must ride without binding. The displacer pushrod (music wire) must pass through the pinhole with as little effort as possible (without creating too much of a leak).

Determine the rotation direction for your engine. This is determined by the 90° offset between the two sections of the crankshaft. The displacer will rise to the top of its stroke a quarter of a turn before the diaphragm is at the top if its stroke. Rotate the flywheel. When you see the displacer moving ahead of the piston, that is the direction in which the engine will run.

You may reduce friction further with some lubrication. Apply a drop of 3-In-One® oil (or another light lubricant) at each point where the crankshaft touches another part.

When you have done your best to seal any leaks and minimize any friction, you are ready for your first test run!

Run the Engine

Place ice or very cold water in the top of the engine. Do not let any water get into the pinhole of the engine, as this may stop the engine from running.

Set the engine on a hot surface, such as the burner of a stove or a hot plate. When using an electric stove, set the burner temperature to low or medium-low. The burner does not need to exceed 300° F in order to operate the engine. The glue in the engine will fail if temperatures climb much beyond this.

Rotate the engine to get it started. These motors will not self start.

Watch the engine as it runs to look for problems that you may need to correct. Sometimes you may need to make some corrections to prevent the crankshaft from wandering out of position or to keep the pushrods in their proper locations.

Troubleshooting

If your engine is not working well, or not running at all, here are some things to check and some hints to help you find success. The key things to observe are **engine timing**, **pressure leaks**, **temperature differential**, and **friction**.

Engine Timing

The power stroke must be 90° behind the motion of the displacer. As the crankshaft rotates and the displacer hits the top of its motion path, the power stroke should be half way up in its rotation. Even if the engine is not running on its own, it should be obvious which way it wants to run. It should at least glide for a few turns in one direction better than it goes in the other. If the timing is not set at 90°, or close to it, this must be repaired.

Pressure Leaks

Pressure leaks can be quite subtle. If you have a pressure leak you will probably hear it hissing as you rotate the crankshaft when heat and ice are present. Tracking that hiss down may prove to be a bit of a challenge.

Try a drop of light oil at the pinhole on the top of the pressure chamber. Watch the oil as you rotate the flywheel while at running temperature. If you see the oil spraying up out of the hole, that is a pressure leak. Often times the presence of the oil is enough to slow the leak and make the engine run. Some pressure has to leak out at this spot. It is impossible to stop it all. The goal here is to minimize the leak enough to let the engine run.

Look carefully at the balloon diaphragm. It should be held against the outside of the pipe elbow with no wrinkles. A wrinkle here will cause a leak. Carefully remove any wrinkles. Add an additional rubber band or a wire wrap around the pipe elbow to secure the diaphragm if needed. Inspect the diaphragm to make sure there are no holes.

Inspect the joint between the pressure chamber and the pipe elbow. If the pipe elbow is not securely fastened to the side of the can, patch it with some additional epoxy.

Increase the Temperature Differential
Use ice on the top of the pressure chamber to maximize the temperature differential. Increase the amount of heat under the pressure chamber to increase the temperature differential.

Friction
Friction may be the most difficult problem to notice. Friction occurs anyplace where two moving parts touch, and anyplace where a moving part contacts the atmosphere.

Sound is caused by friction, but not all friction creates a sound you can hear. So if you can hear an obvious grinding, scraping, or squeaking sound, you probably need to fix it.

It is sometimes necessary to make adjustments to the mechanical components of the crank mechanism to reduce friction. Check every point that touches the crankshaft to see if there is a problem. Here are some of the potential problems and how to fix them:

- Does the crankshaft drift to one side and bind up when a bend in the shaft gets close to the side of the can? This can be compensated for in a couple of ways. One is to level the crankshaft by tilting the engine slightly. The other method is to wrap a small piece of wire, or place a small piece of coffee straw, around the crankshaft to prevent it from drifting too far out of alignment. This collar will prevent the shaft from drifting and touching the side of the can.
- Is the diaphragm pushrod perpendicular to the axis of the crankshaft, or is it pushing up at an angle other than 90°? The diaphragm pushrod needs to be going straight up and down. If it is pushing at an angle other than 90°, it may be binding and must be repaired. You will have to do some troubleshooting here to find what needs to be done.
- If a section of the crankshaft is not parallel to the axis of rotation, that can cause a problem. This is repaired by correcting the bend in the crankshaft.
- Friction can also be caused by not having the foot of the pushrod centered on the diaphragm. Fixing the problem may be as simple as bending the pushrod or re-centering the diaphragm.
- Another way to fix a wandering pushrod is to wrap some wire around the crankshaft to prevent the pushrod from wandering out of alignment.
- Check the travel of the displacer to make sure it moves freely and does not impact the top or bottom of the pressure chamber. Make any adjustments necessary to make sure the displacer can move freely as the crankshaft rotates. If the displacer hits both the top and bottom of the pressure chamber, adjust the crankshaft to reduce the amount of travel.

Look for these titles and more by Jim R. Larsen:

THREE LTD
STIRLING ENGINES
YOU CAN BUILD

WITHOUT A MACHINE SHOP

JIM R. LARSEN

This book will guide you step-by-step through the process of building three low temperature differential (LTD) Stirling Engines for less than $30 each. Two of them are efficient enough to run from the heat of your hand. And all three of them can be built using common hand tools with material available at your local hardware store and hobby shop.

The hot air engine (aka: "Stirling engine") has been around for a very long time. Interest in the technology has been revived in recent years by those exploring new methods for producing "Green Energy". An LTD Stirling engine is efficient enough to run from the waste heat that is discharged from other appliances or from the warmth of direct sunlight. The most efficient of these will run from the heat of your hand.

Until now, if you wanted to build your own heat of the hand Stirling Engine you had to either spend several hundred dollars for an expensive kit or you needed access to a precision machine shop. This book breaks both those barriers for you. Now you can build your own working models that will run from the heat of your hand, without the aid of a machine shop.

Whether you are a student looking for a winning science fair project, or just a home hobbyist looking for a fun project, you will find this book to be a helpful guide in creating your own LTD Stirling engines. You will find detailed instructions, tips on finding parts, and over 130 illustrations to help you recreate Jim Larsen's original designs for these highly efficient small models

ELEVEN STIRLING ENGINE PROJECTS

YOU CAN BUILD

JIM R. LARSEN

Here is a collection of eleven Stirling engine projects, including 5 new groundbreaking designs by Jim Larsen. Now you can build simple pop can Stirling engines that look sharp and run incredibly well. Four of the projects are air cooled pop can engines that will run for hours over a simple candle flame.

The collection provides complete illustrated instructions for building the Single Chamber Pop Can Stirling Engine, Dual Chamber Pop Can Stirling Engine, Walking Beam Pop Can Stirling Engine, Horizontal Pop Can Stirling Engine and the Quick and Easy Stirling Engine.

Kit builders will enjoy the detailed reviews of 4 commercially available kits: Thames and Kosmos Stirling Engine Car and Experiment Kit, Think Geek Stirling Engine Kit by Inpro Solar, MM5 Coffee Cup Stirling Engine Kit by the American Stirling Company, and the Grizzly H8102 Stirling Engine Machined Kit.

Also included are these classic designs: The SFA Stirling Engine Project (Stephen F. Austin University), and the Easy to Build Stirling Engine (Geocities/TheRecentPast).

Visit http://StirlingBuilder.com for more help and support with your Stirling engine projects.

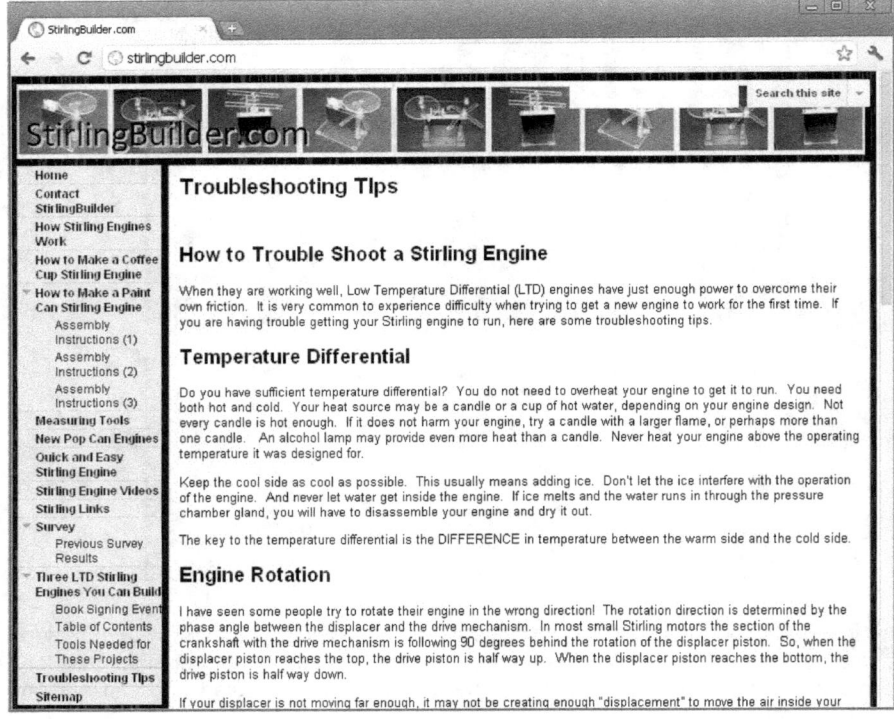

The StirlingBuilder website contains free Stirling engine designs and links to helpful videos and articles. Simply search for StirlingBuilder with your favorite search engine.

www.ingramcontent.com/pod-product-compliance
Lightning Source LLC
Chambersburg PA
CBHW051253170526
45165CB00004B/1693